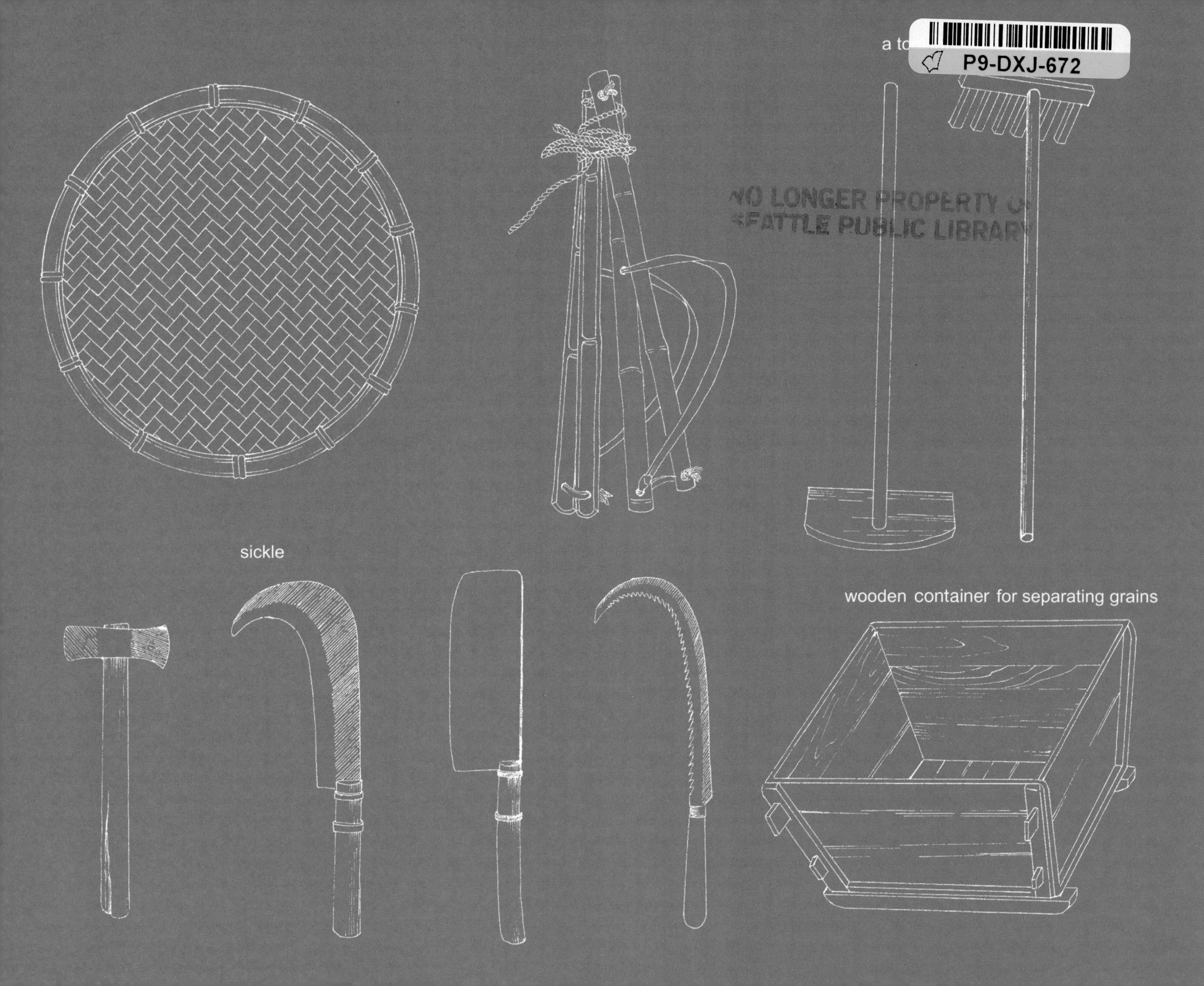

P9-DXJ-672
NO LONGER PROPERTY OF
SEATTLE PUBLIC LIBRARY

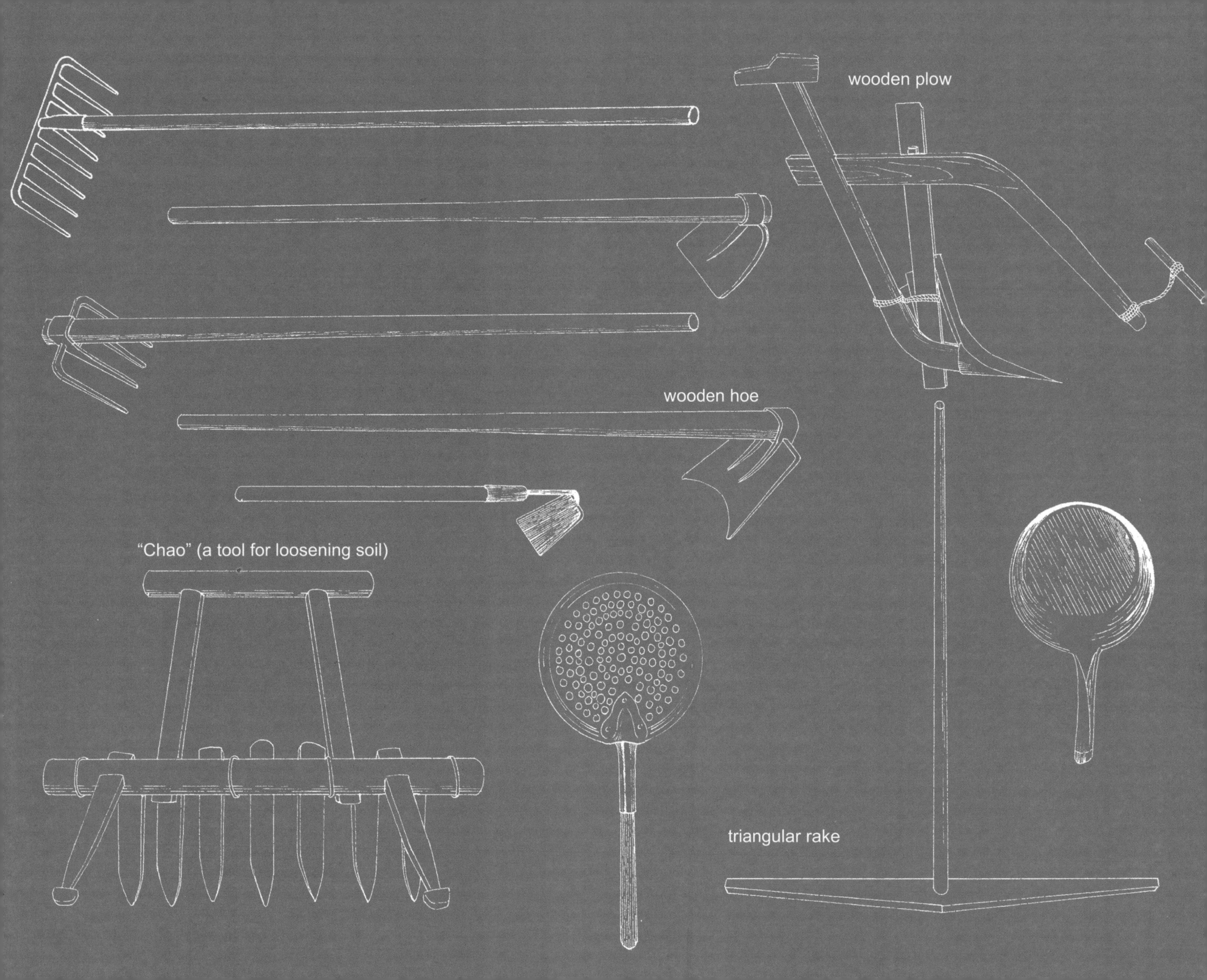
wooden plow
wooden hoe
"Chao" (a tool for loosening soil)
triangular rake

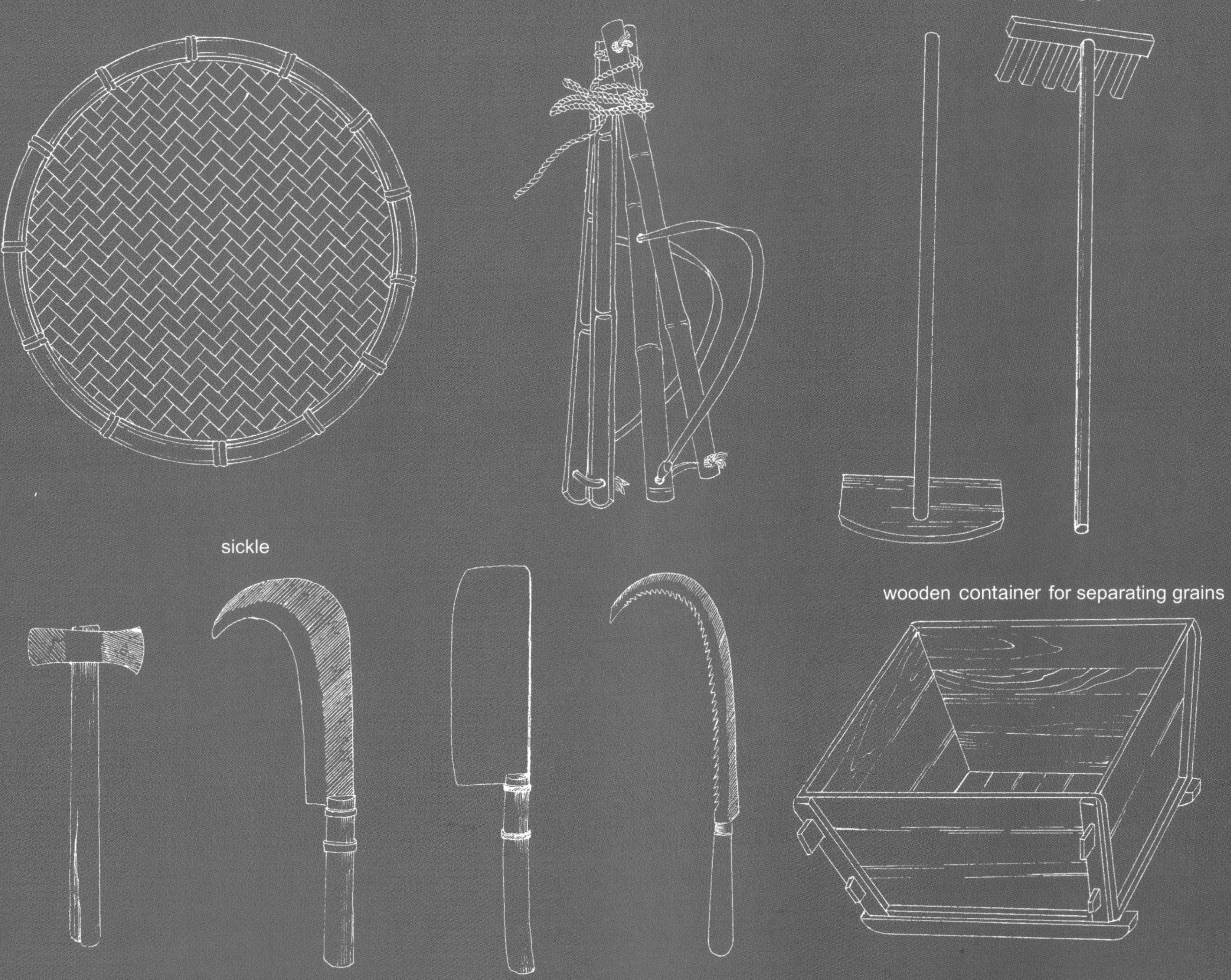
a tool for spreading grain
sickle
wooden container for separating grains

Reycraft Books
55 Fifth Avenue
New York, NY 10003

Reycraftbooks.com

Reycraft Books is a trade imprint and trademark of Newmark Learning, LLC.

This edition is published by arrangement with China Children's Press & Publication Group, China.
© China Children's Press & Publication Group

All rights reserved. No portion of this book may be reproduced, stored in a retrieval system, or transmitted in any form or by any means, electronic, mechanical, photocopying, recording, or otherwise, without written permission from the publisher. For information regarding permission, please contact info@reycraftbooks.com.

Educators and Librarians: Our books may be purchased in bulk for promotional, educational, or business use. Please contact sales@reycraftbooks.com.

This is a work of fiction. Names, characters, places, dialogue, and incidents described are either the product of the author's imagination or are used fictitiously. Any resemblance to actual persons, living or dead, is entirely coincidental.

Sale of this book without a front cover or jacket may be unauthorized. If this book is coverless, it may have been reported to the publisher as "unsold or destroyed" and may have deprived the author and publisher of payment.
Library of Congress Cataloging-in-Publication Data is available.

ISBN: 978-1-4788-6936-8

Printed in Guangzhou, China
4401/1119/CA21902021
10 9 8 7 6 5 4 3 2 1

First Edition Hardcover published by Reycraft Books 2020.

Reycraft Books and Newmark Learning, LLC, support diversity and the First Amendment, and celebrate the right to read.

Rice

by Hongcheng Yu

On the terraces in the beautiful mountains of Yunnan Province, local farmers grow rice. After the rains, the terraces are warm and wet. As the cherries blossom, people start preparing to work on the farm.

The Rains
February 18–20
As winter passes and spring draws near, the weather gets warmer and warmer.

Before planting, farmers have to plow the soil, which has hardened during the winter. They use tools like a plow and a rake to soften the earth so that the rice seeds can root easily.

The Rains
February 18–20
As winter passes and spring draws near, the weather gets warmer and warmer.

New leaves sprout from the trees and animals become active. This is the time that farmers begin to prepare their seeds for sowing. They choose big, strong seeds, soak them in warm water, and leave them in a warm and wet bamboo basket for three to four days, until the seeds begin to sprout. The hen and the chicks eat any seeds that have dropped on the ground.

Insects Awaken
March 5–7
Thunderstorms arrive.

The flowers on the pear and peach trees blossom and wilt. Farmers begin to plant the sprouted seeds into flat, soft seedbeds. In the warm sunlight, the seedlings will become healthy and strong.

Insects Awaken
March 5-7
Thunderstorms arrive.

It drizzles thick and fast on the Pure Brightness days. Potatoes start to blossom, and a grandma plows and plants her corn. The rice seedlings begin to form their first three leaves. Farmers go to the seedbed regularly to check on their seedlings.

Qingming—Pure Brightness
April 4-6
Everything grows in the sun.

By now there are five to six leaves on the rice seedlings, which makes the seedbed a very crowded place! It's time to take the seedlings out and carry them to the rice paddy. Care has been taken to make sure the paddy has been plowed and the dam reinforced.

Summer Begins
May 5–7
Summer arrives, and everything thrives.

It takes a lot of effort to transplant the seedlings, so all the neighbors and relatives come to help. All the seedlings are transplanted securely in rows in the rice paddy. After an entire day of hard work, the transplanting is finished for only one family. They will have to do the same work in their neighbor's field the next morning. Several days later, the hills will be covered with endless greenery.

Summer Begins
May 5–7
Summer arrives, and everything thrives.

The leaves of the rice plants rustle when the rain whispers to them, "Little grain, may you prosper and bring everyone a bountiful harvest." As the seedlings adapt to their new environment, they become stronger and stronger. Farm children play on the swing near the field, hoping to drive bad weather and pest problems away as they fly up and down.

Grain Full
May 20–22
It does not rain on Grain Full day. It is not wet on Grain in Ear day.

Small flowers bloom on the ears of rice. There will be two to three hundred of these flowers, which will then be pollinated and grow into grains. The plants have to absorb a lot of water for this to happen, which means a strong rain in this hot weather is desperately needed.

Great Heat
July 22–24
If it is hot during Great Heat, let's have a rest. If it is cool, water fills the pool.

The big and heavy ears make the rice stalks bow. To prevent them from falling into the water and drowning during storms, they are bundled up tightly. As the rain stops and the sky becomes clear, sparrows will peck at the rice grains. Scarecrows are placed in the field to scare them away.

Autumn Begins
August 7–9
The autumn sky is clear and the air is crisp.
The moon is bright and the wind is light.

Now the grains are completely mature. On a sunny day, the family harvests the rice together. The ears of rice are cut with a sickle and then put into a wooden container so the grains can be separated. Farmers separate the grains by shaking them inside the container. The rattling sound echoes through the village. Carps, loaches, and finless eels have grown big in the paddies. Children catch these fish after the harvest. They make a delicious dinner!

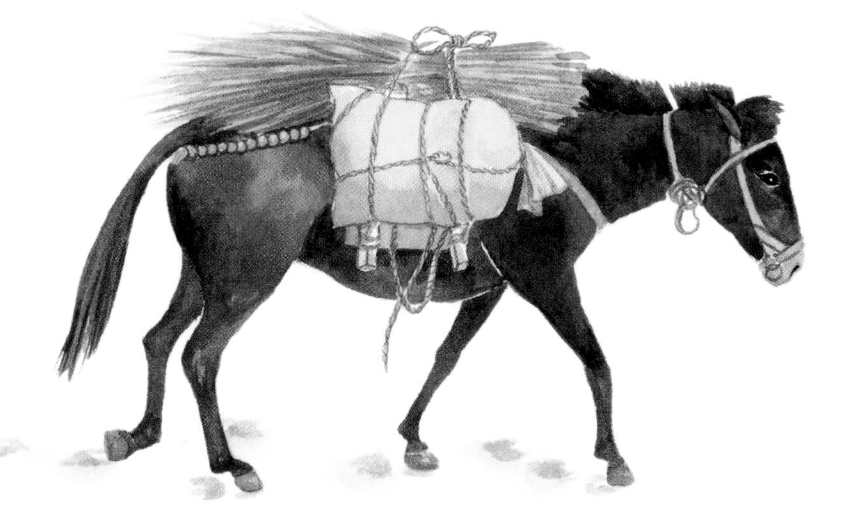

Autumn Equinox
September 22–24
The nights of the White Dew and Autumn Equinox are colder than previous nights.

In order to preserve the grains of rice, people dry them in the sun. On sunny days, families spread them on flat ground near their homes, turning them over regularly with a special tool to ensure that every grain dries.

Cold Dew
October 8–9
The air is cold, and the dew begins to freeze.

There are husks on the outside of the rice grains, which must be cleaned so the rice can be eaten. For many years, farmers used water-powered rollers to mill, or husk, the grains. Now a machine blows the husks away. At last, they have rice! As the biggest festival in October draws near, every corner of the village is cleaned and the best food is prepared. A banquet is held in the streets to celebrate a bountiful year and to pray for good luck in the coming year.

Frost's Descent
October 23–24
Winter begins, and dew becomes frost.

NOTES

Ancient Chinese people used a lunar calendar based on the various cycles of the moon. Since it is different from the solar calendar used today, it didn't always correctly reflect the change of seasons. Ancient astronomers divided the zodiac into 24 equal segments based on the position of the sun and came up with 24 solar terms. These terms predict changes in weather, water, and frost, and are important benchmarks in farming.

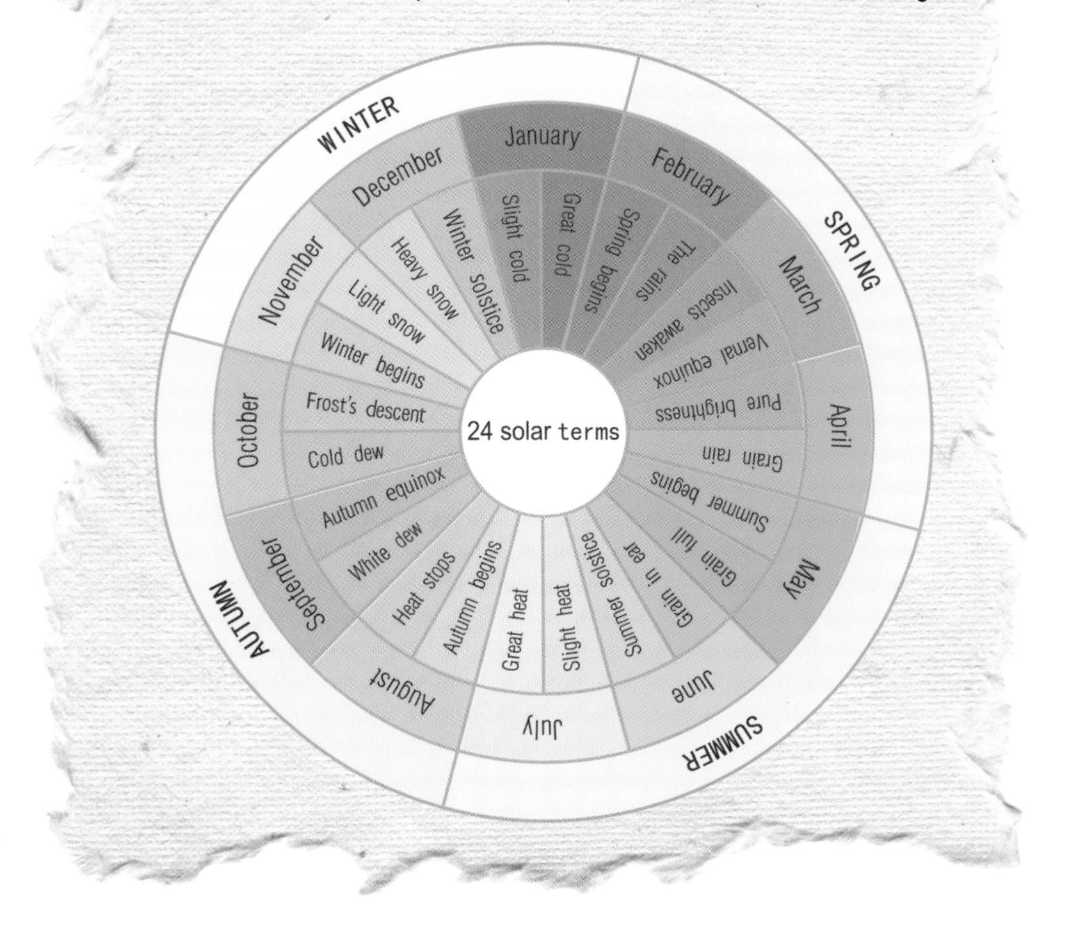

Pages 4–5

The Rains

There are many mountains and hills in southwest China. In order to grow rice in these areas, people build terraces. Dams help to store spring water. The overflow from the upper terraces fills the lower ones, and eventually all the fields, from halfway up the mountain to the valley, are filled with water.

Pages 6–7

The Rains

The work starts with making a seedbed. After winter, the soil has become hard and dry. Farmers plough the soil, break it up with a "chao," soften it with water, and smooth it with a rake. Finally, the soft and wet soil is a good place for the rice seeds to root.

Wooden plow

As the sharp end of the V-shaped plow is inserted into the soil, oxen start to pull the plow. Doing this buries old stems and roots, which then turn into fertilizer.

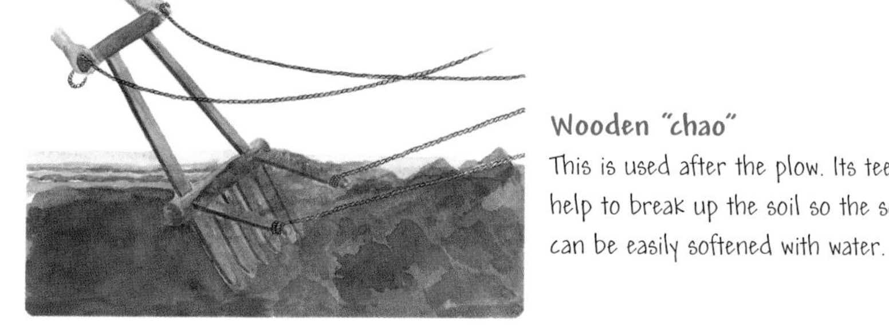

Wooden "chao"

This is used after the plow. Its teeth help to break up the soil so the soil can be easily softened with water.

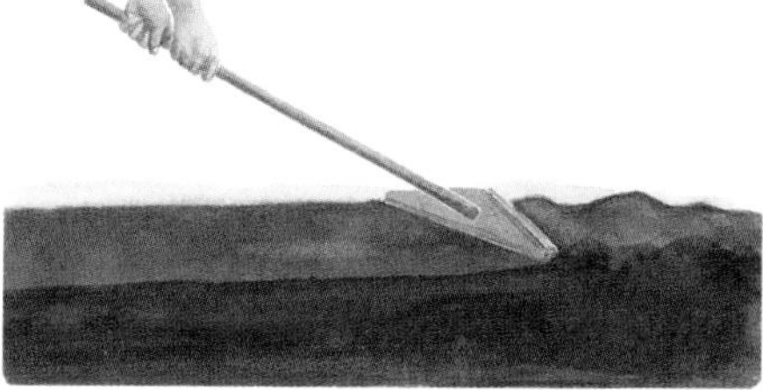

Triangular rake

This helps smooth the soil so that the rice can be properly distributed.

Pages 8–9

Insects Awaken

At this time, seeds are prepared, The biggest, healthiest seeds are chosen and soaked in 35°C (95°F) water. The absorbed water amounts to 40% of a seed's weight. The seeds expand, soften, and start to germinate.

1. Remove withered seeds, straw, and other debris.

2. Put the seeds in a container and pour warm water over them. Remove the floating ones that might be damaged.

3. Add more warm water and soak the seeds overnight.

4. Use warm water to clean the seeds and then put the seeds in a basket covered with straw.

5. Cover the straw-packed seeds with stones, which will help maintain the seeds' temperature and humidity.

6. Water the seeds every day with 35°C water. Excess water will simply seep though the bamboo basket.

Pages 10–11

Insects Awaken

Seeds will germinate in the warm basket after three to four days. Then they can be planted in the seedbed.

Day 1.

Day 2. Seeds start to germinate.

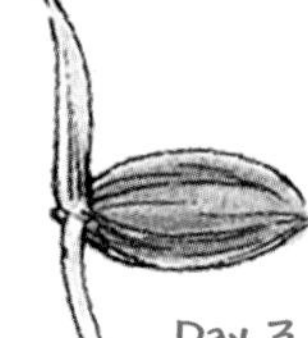

Day 3. When the buds are half as long and the roots are as long as the seeds, they can be planted in the seedbed.

With enough sunlight, water, and air, the seeds can grow well. The seedbed should be surrounded by small ditches, so that excess water can flow down into the ditches. This will help keep the soil moist without drowning the seeds.

Pages 12–13

Pure Brightness

After twenty days in the seedbed, the seed starts to grow its first three leaves. Prior to this, nutrition came from the seed itself. But at this point, the seed starts absorbing water and nutrients from the soil through its roots and uses photosynthesis to acquire nutrients through its leaves. Seedlings are easily damaged, so farmers will need to drain off excess water in the seedbed.

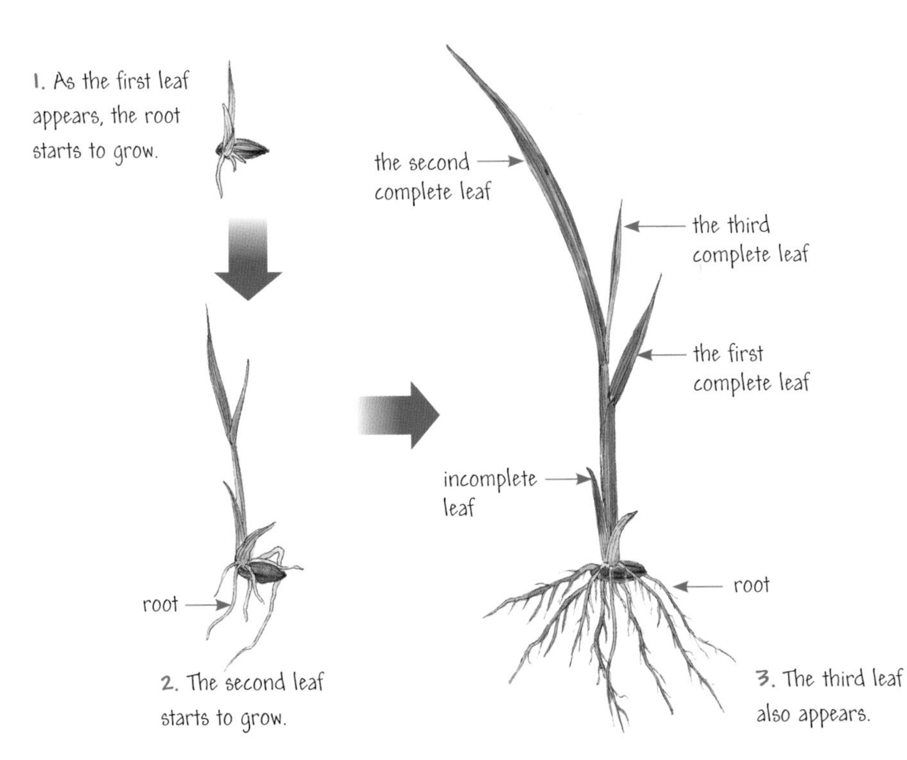

Pages 14–15

Summer Begins

As the fifth and sixth leaves begin to grow, it gets quite crowded in the seedbed. It's time to take the seedlings out and transplant them into the paddy field.

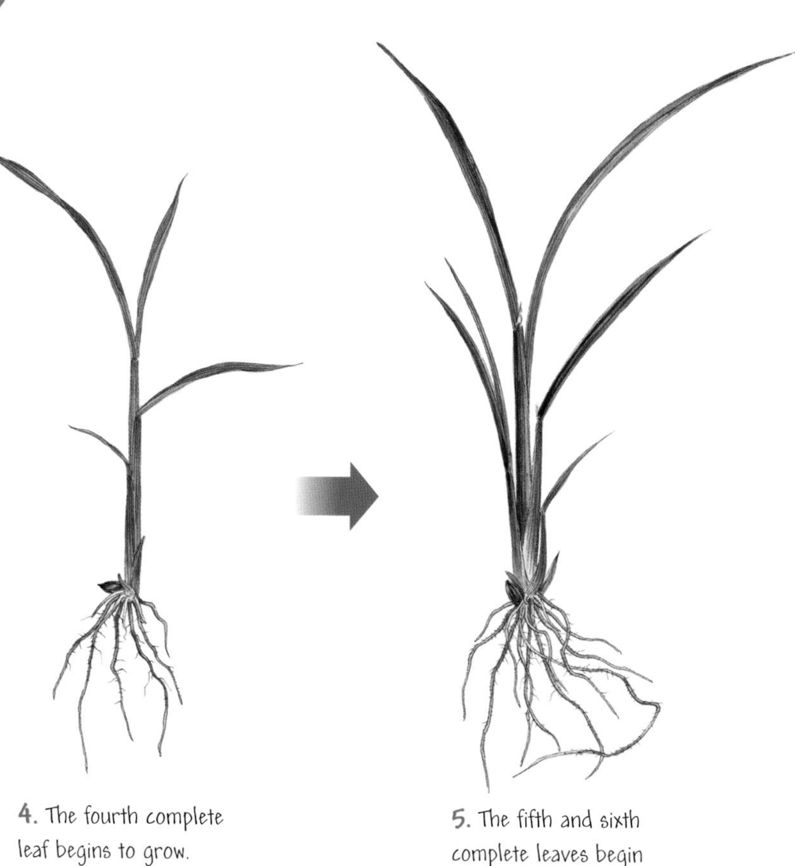

4. The fourth complete leaf begins to grow.

5. The fifth and sixth complete leaves begin to grow.

Pages 16–17

Summer Begins

It is very crowded in the seedbed. When seedlings are transplanted, there should be some space left between them so they have room to grow.
The seedlings should be planted two centimeters deep in the paddy soil to prevent them from floating. Farmers often fill holes with their feet when they back up, and then plant the seedlings. The water reaches two-thirds of the way up the seedlings, which helps them recover from the damage of transplantation.

Pages 18–19

小满

Grain Full

As the roots grow, the seedlings start to tiller (which means to put forth new shoots from the stalk). The more the seedlings tiller, the more ears they gain. People often keep the water as deep as three to five centimeters to promote tillering. Ears begin to grow from the stalk after the seedlings tiller several times, during which time the farmers weed and remove pests.

First tillering:
new shoots grow from the stem

Second tillering:
new shoots grow from the shoots of the first tillering

Pages 20–21

大暑

Great Heat

As the new ears grow, they start to bloom. The flowers have stamens and lemmas rather than petals. The blooming starts at the top and lasts for one week.

Rice flowers are self-pollinating. When they bloom, the pollen from the anther at the end of the stamen falls on the pistil. After pollination, the seed starts to grow.

Pages 22–23

Autumn Begins

After blooming and pollination, the seed grows gradually in the glumes. Sparrows often like to peck at the grains.

1. Three to five days after blooming, the endosperm develops, which makes the ovary grow. The nutrition comes from the stem and the leaves.

2. The endosperm becomes thickened and waxy. The hull becomes yellow.

3. Seven or eight days later, the endosperm becomes heavy, which makes the rice bow.

Pages 24–25

秋分

Autumn Equinox

It takes great effort to harvest the grains, yet farmers happily do the hard work. There are a few steps to harvesting.

1. Sickle

The sickle is moon-shaped and has teeth on its sharp edge. Farmers use one hand to hold the stalks and the other to cut them.

2. Wooden container for separating grains

This is an ancient tool for separating rice grains. Because it's lightweight, it's very useful in mountainous regions. It can even float on water. Farmers separate the grains by shaking them inside the container.

3. Packing

Farmers use a small bamboo basket to put the grains in different bags. Then they are distributed to families by human or horse.

4. Straw

The straw is bundled up and drained, and can be used as oven fodder, fuel, heat preservation, or house-building material. It's quite useful!

Pages 26–27

Cold Dew

Drying the grains in the sun preserves them for easy husking and transportation.

1. It's important that this is done on a sunny day.

2. Spread the grains on a flat area or nearby platform with the special tool to keep them dry and warm.

3. Use the tool to turn them over regularly.

4. Remove straw, withered grains, and other debris. Store the grains in bags.

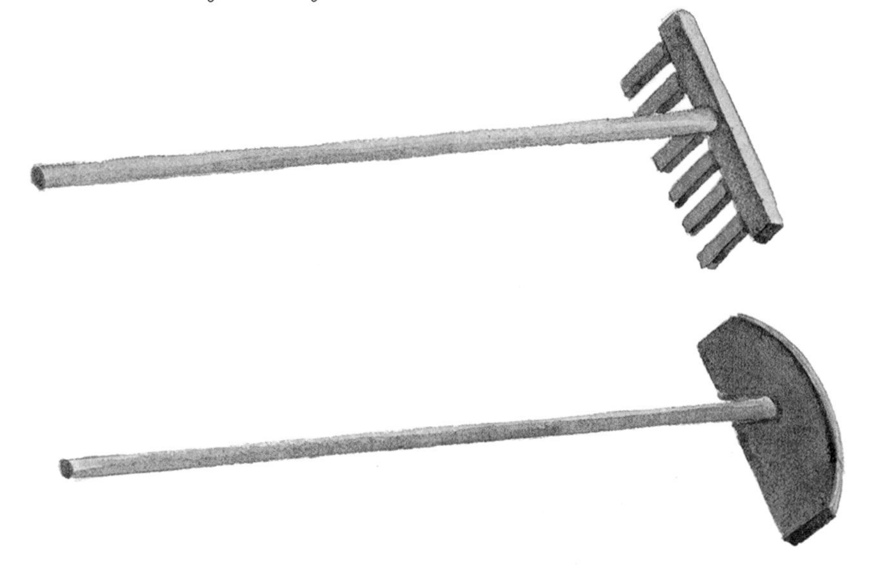

Pages 28–29

霜降

Frost's Descent

People use different tools to mill rice and polish the grains.

1. Water-powered roller This milling tool was invented sometime during the Wei, Jin, or Northern and Southern Dynasties and is powered by water and made of stones. Waterwheel A is powered by water and drives bearing B and gear C. Gear C turns gear D, causing bearing E to rotate. Through another bearing, F, G can rotate in the stone groove H. Since G and H have cracks and bulges, husks can be easily removed when the grains are between them. The rice can be preserved since the sun-dried grains are relatively hard.

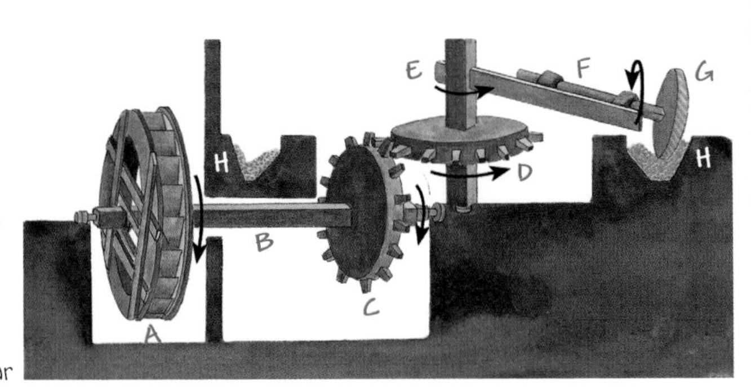

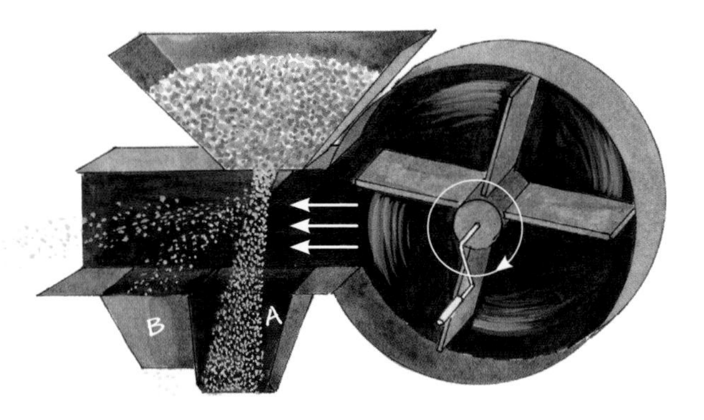

2. Husk-blowing roller machine It's the traditional machine used in southern China to remove straw, withered seeds, and other debris from rice. Pour the milled grains into the funnel. Roll the fan blade roller. The lighter husks are blown off first, while the heavier withered and broken grains fall into funnel B. Those in the middle, the full grains, fall into funnel A.

Why do we eat rice?

Rice is a staple food in China. It gives people energy and nutrients that vegetables and meat don't provide.

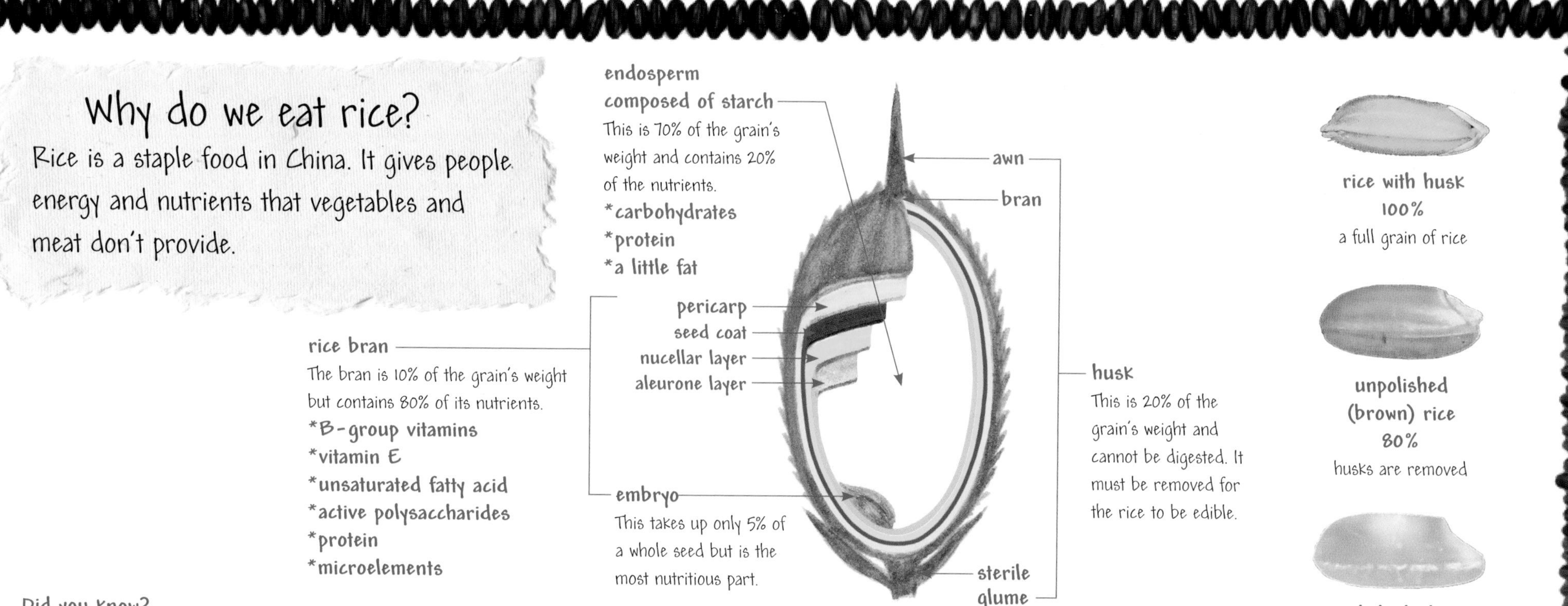

Did you know?

There are more than 140,000 species of rice, which can be roughly divided into two types: japonica rice and indica rice. They can also be divided into another two types—glutinous and non-glutinous. Let's have a look at them. Do you know the differences between them? Which one do you regularly eat?

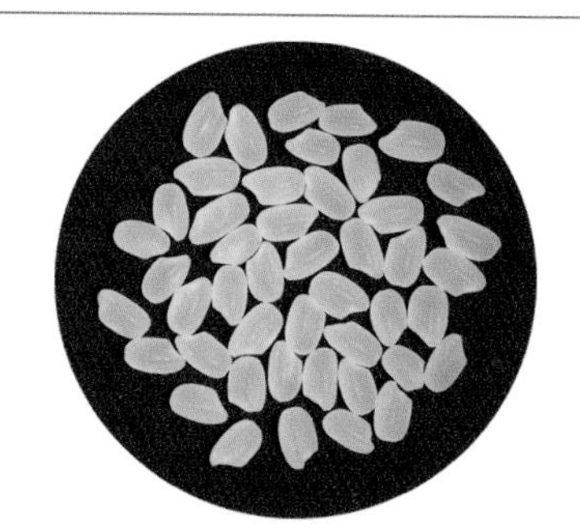

Japonica (non-glutinous) rice
These rice grains are short, round, white, relatively translucent, and not very glutinous. This type of rice is very common. It includes northeastern Chinese rice, Japanese sushi rice, etc.

Indica (non-glutinous) rice
These rice grains are thin, long, white, relatively translucent, and not glutinous. This is also a popular type of rice. It includes Chinese silver needle rice, Thai hom rice, etc.

Japonica (glutinous) rice
These grains are short, round, and white. Most of them are not translucent. This rice contains almost 100% amylopectin and is very glutinous. It is used in brewing and to cook yuanxiao, which is a kind of sweet dumpling eaten by Chinese people during their Spring Festival.

Indica (glutinous) rice
This rice is thin, long, and white. Most of it is not translucent. It is very glutinous. People usually use it to make shaomai and vongzi.

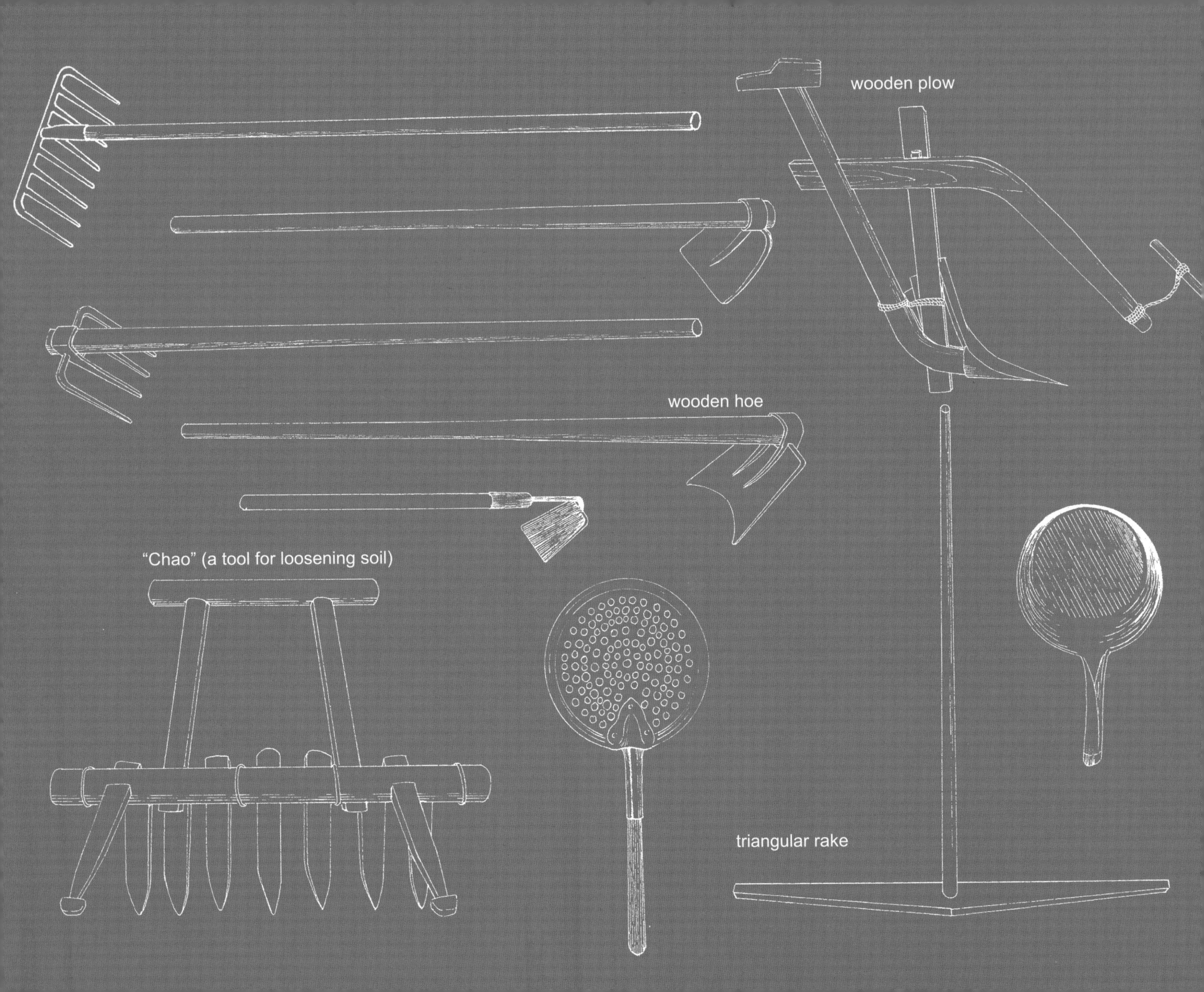
wooden plow
wooden hoe
"Chao" (a tool for loosening soil)
triangular rake

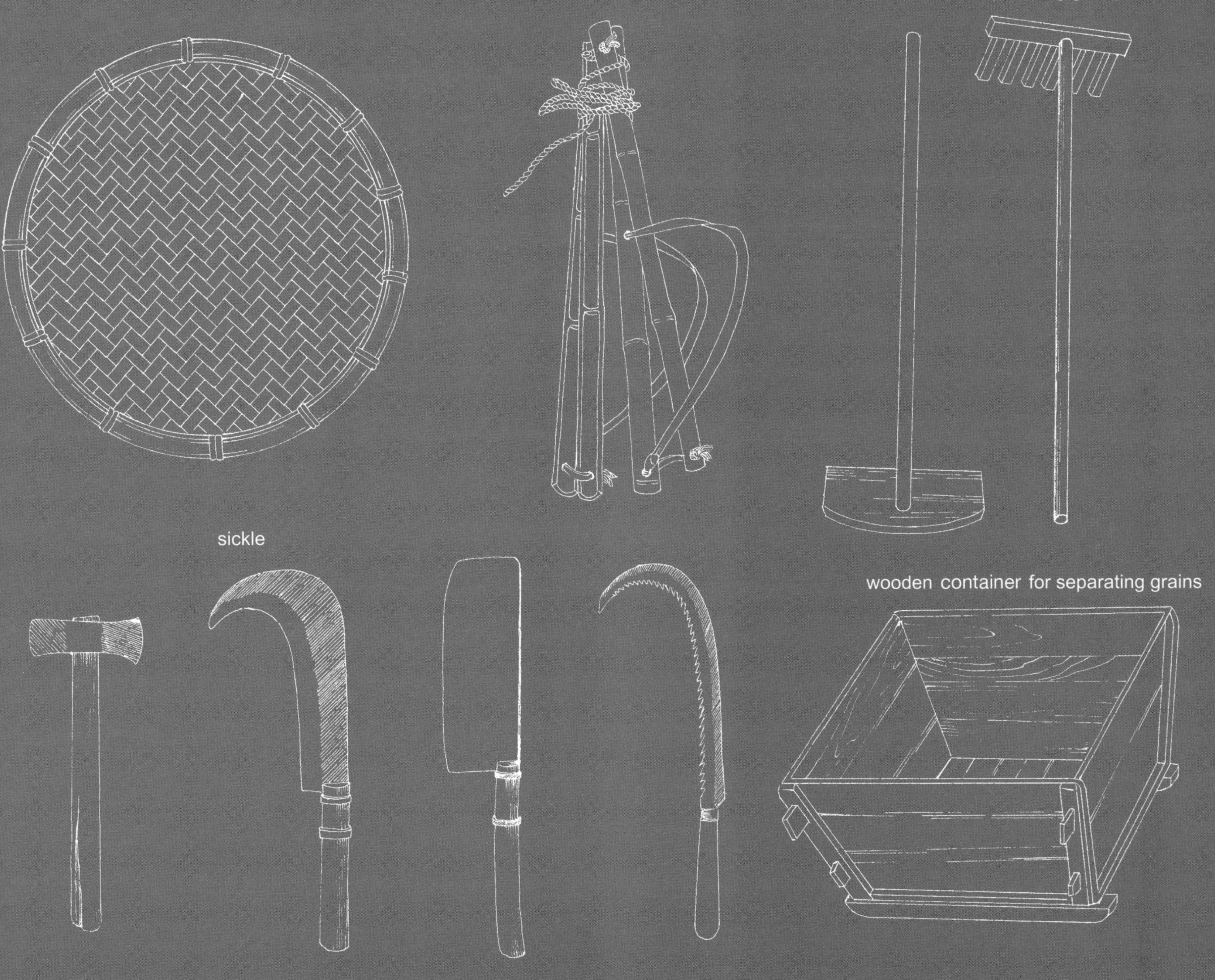
a tool for spreading grain
sickle
wooden container for separating grains

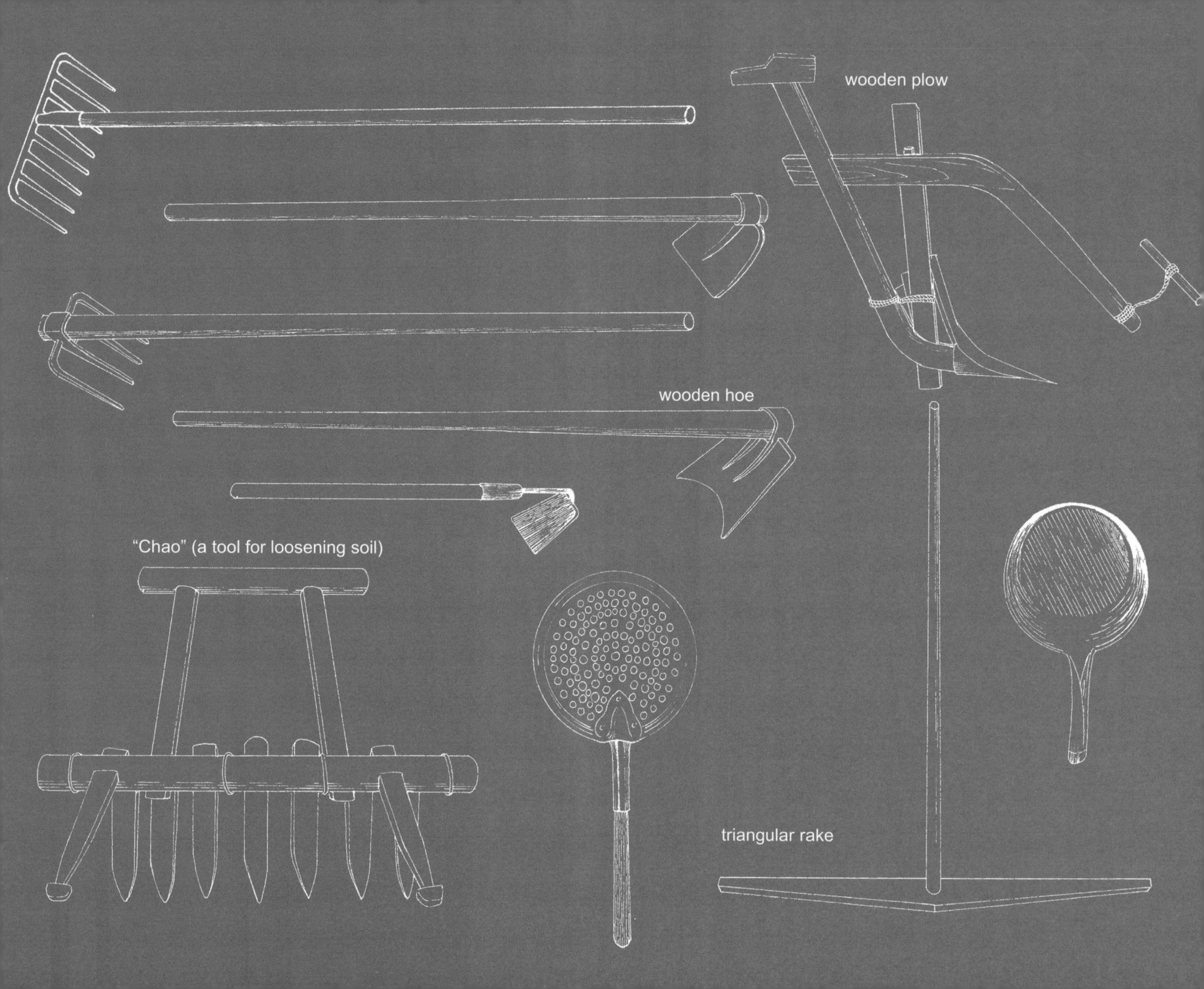
wooden plow
wooden hoe
"Chao" (a tool for loosening soil)
triangular rake